WORKING WITH MACHINES

Sonya Newland

Kane Miller
A DIVISION OF EDC PUBLISHING

First American Edition 2022
Kane Miller, A Division of EDC Publishing

Editor: Sonya Newland
Illustrator: Diego Vaisberg
Designer: Clare Nicholas

First published in Great Britain in 2020 by Wayland, An imprint of Hachette Children's Group, Part of The Watts Publishing Group, Carmelite House, 50 Victoria Embankment, London EC4Y 0DZ

For information contact:
Kane Miller, A Division of EDC Publishing
5402 S 122nd E Ave, Tulsa, OK 74146
www.kanemiller.com
www.myubam.com

Library of Congress Control Number: 2021937030

Printed and bound in China
1 2 3 4 5 6 7 8 9 10

ISBN: 978-1-68464-331-8

All the materials required for the projects in this book are available online, or from craft or hardware stores. Adult supervision should be provided when working on these projects.

CONTENTS

MECHANICAL ENGINEERING

Engineers come up with ideas to solve problems. They invent, design, develop, and build all sorts of things, from smartphones and satellite navigation systems to skyscrapers and spacecraft. Mechanical engineers work with machines.

What are machines?

A machine is basically any device that does work. That includes everything from a spade or garden shears...

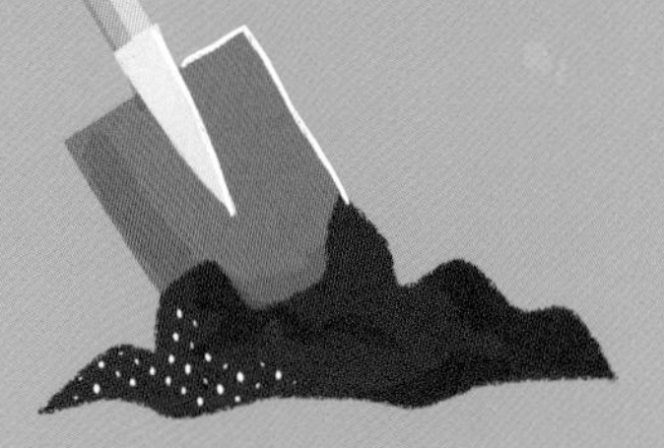

...to an electric kettle or a blender...

...to a fighter jet!

Most machines are designed to help people do things better, more quickly, or more easil

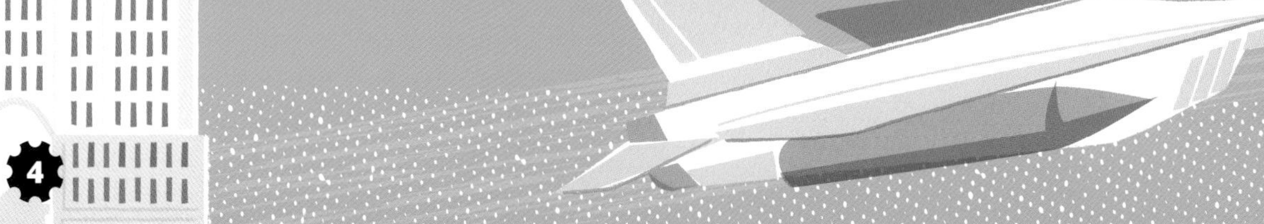

Ancient engineers

Mechanical engineering is one of the oldest forms of engineering. Ingenious engineers in ancient Egypt and ancient Greece invented machines to help with work thousands of years ago. For example, the shaduf was a machine that raised water from a well or river.

The pole could rotate on a beam to move the bucket around.

The bucket was lowered into the well by pulling the rope.

A weight on the other end, called a counterweight, raised the bucket.

The area of **science** related to **machines,** force, and moving parts is called **"mechanics"** so engineers who work with machines are called **"mechanical engineers."**

Could you be an engineer?

Today, mechanical engineers use math and science to understand how things move and work. They apply their problem-solving skills to designing new machines or improving ones that already exist, such as creating a faster, safer bicycle. A mechanical engineer:

identifies a need → comes up with an idea → designs it → builds it → tests it → improves it.

WORK, FORCE, AND MOVEMENT

To design and build the best machines, engineers have to understand two important scientific ideas – force and motion (movement). Force is the amount of energy applied to an object. We use machines to turn force into movement.

How much work?

Work is the amount of energy it takes to move an object over a distance. Engineers use an equation to measure work:

work = force x distance

Making life easier

Machines allow people to do the same amount of work, but with less effort. For example, a machine called a pulley makes lifting loads easier. The amount of work a machine saves is called "mechanical advantage."

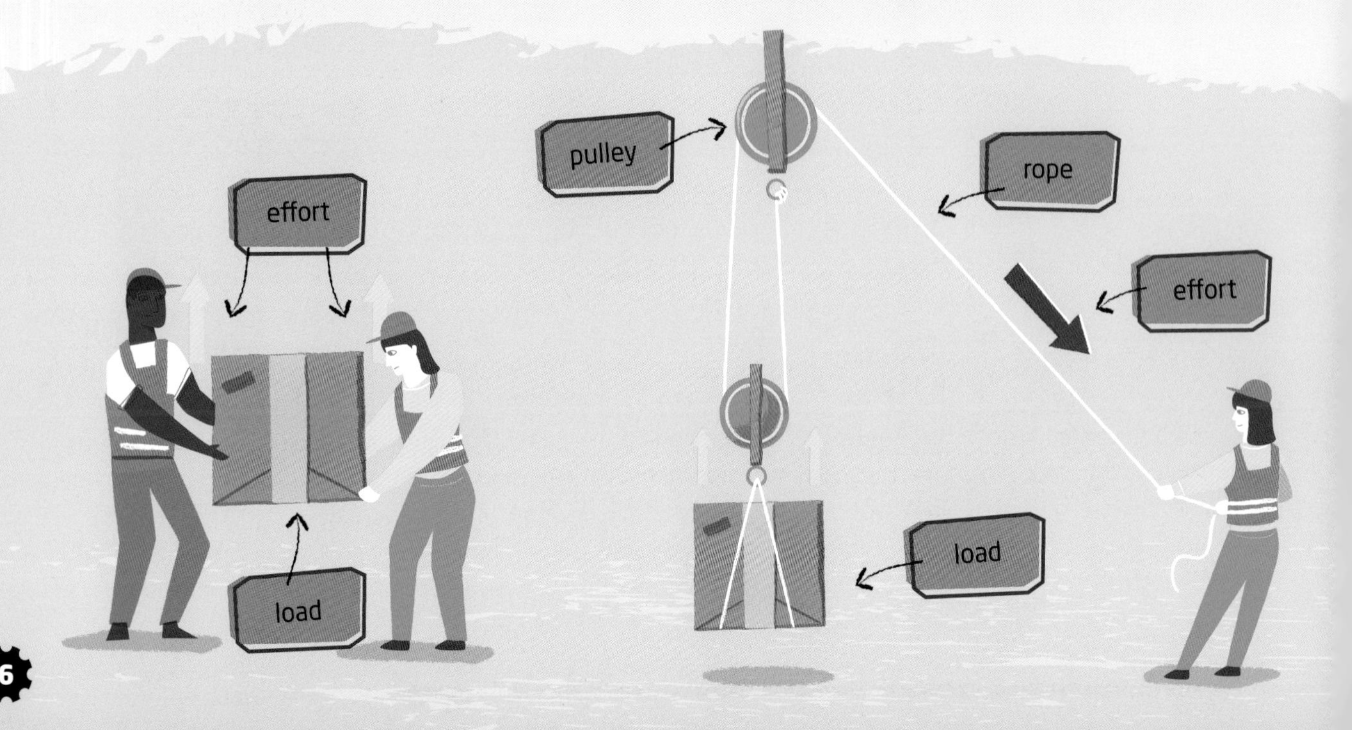

Input force is the force you apply to the **machine. Output force** is the force that the machine applies to the **object.**

Many machines

Engineers design machines that make work easier in different ways.

1. They can increase the amount (magnitude) of force applied.

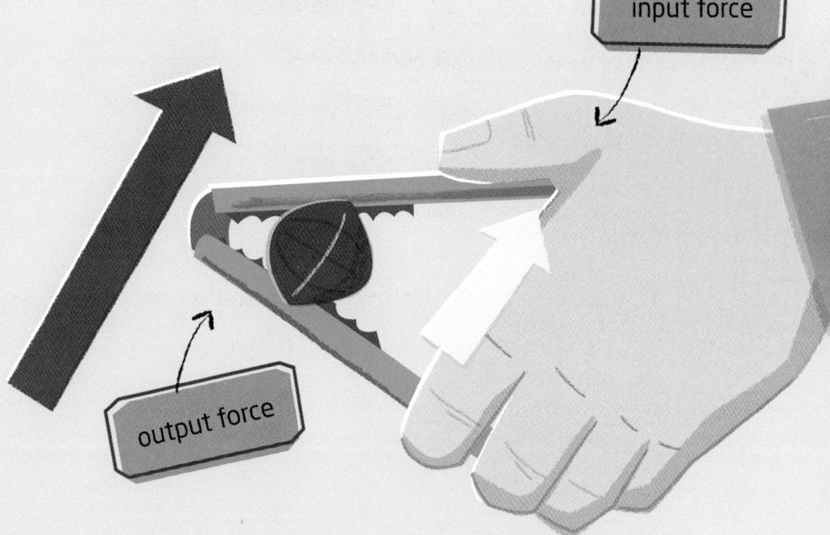

2. They can increase the distance an object moves.

3. They can change the direction in which an object moves.

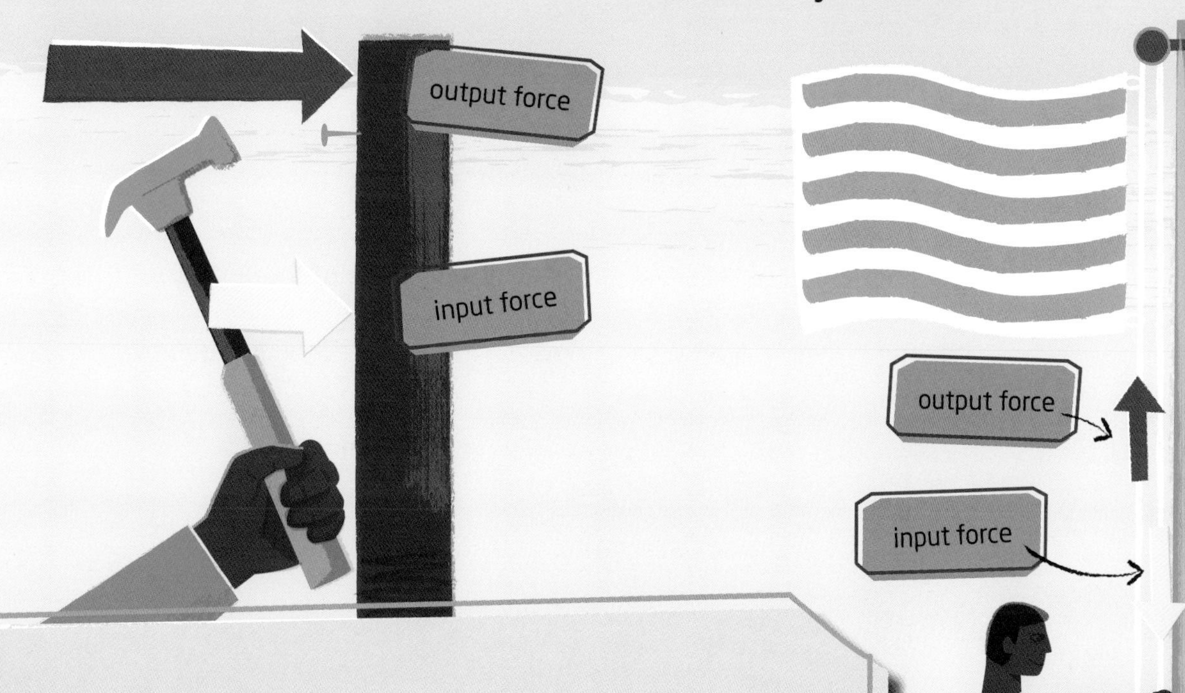

Simple machines

Engineers use six main types of simple machines: lever, pulley, wheel and axle, inclined plane, wedge, and screw. These machines are the basis of all mechanics – and the building blocks of much more complex machines. Read on to find out more about these simple machines.

LEVERS

Engineers use levers in the design of many different machines. You can see them in everything from simple household items to huge, complex machines in factories and on construction sites.

A simple lever

A lever is a flat beam that rests on a pivot or fulcrum.

Engineers use levers to change the direction of a force. Pushing *down* on one end of the lever (input force) lifts *up* the load at the other end (output force). The output force is also bigger than the input force (see page 7).

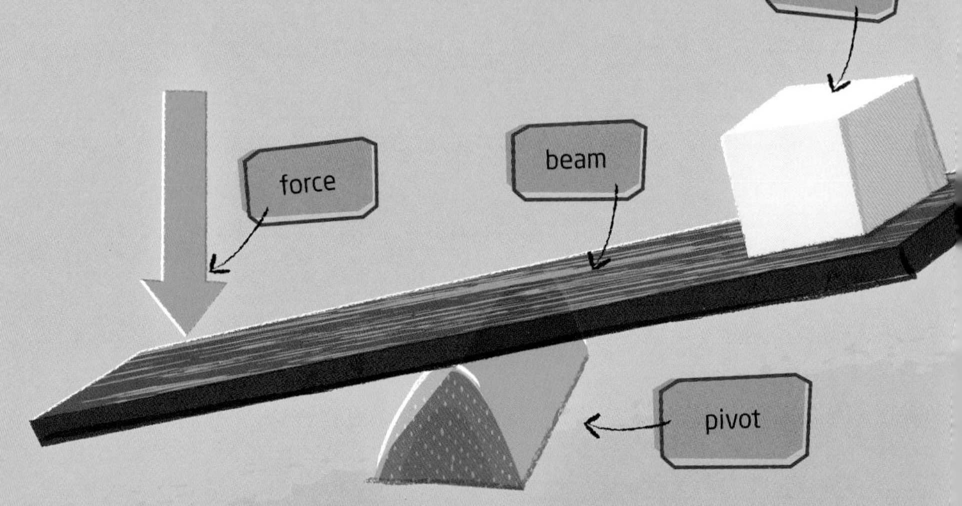

Three classes of lever

The parts of a lever can be positioned in three different ways. Engineers refer to these as three "classes" of lever.

Class 1
The pivot is between the force and the load. The closer the load is to the pivot, the bigger the mechanical advantage.

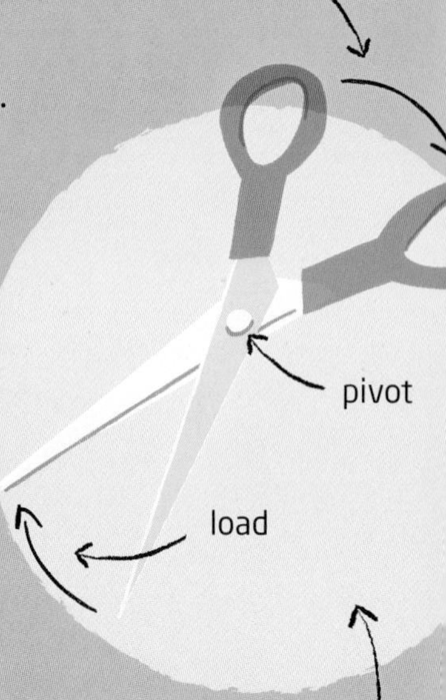

Scissors are an example a class 1 lever.

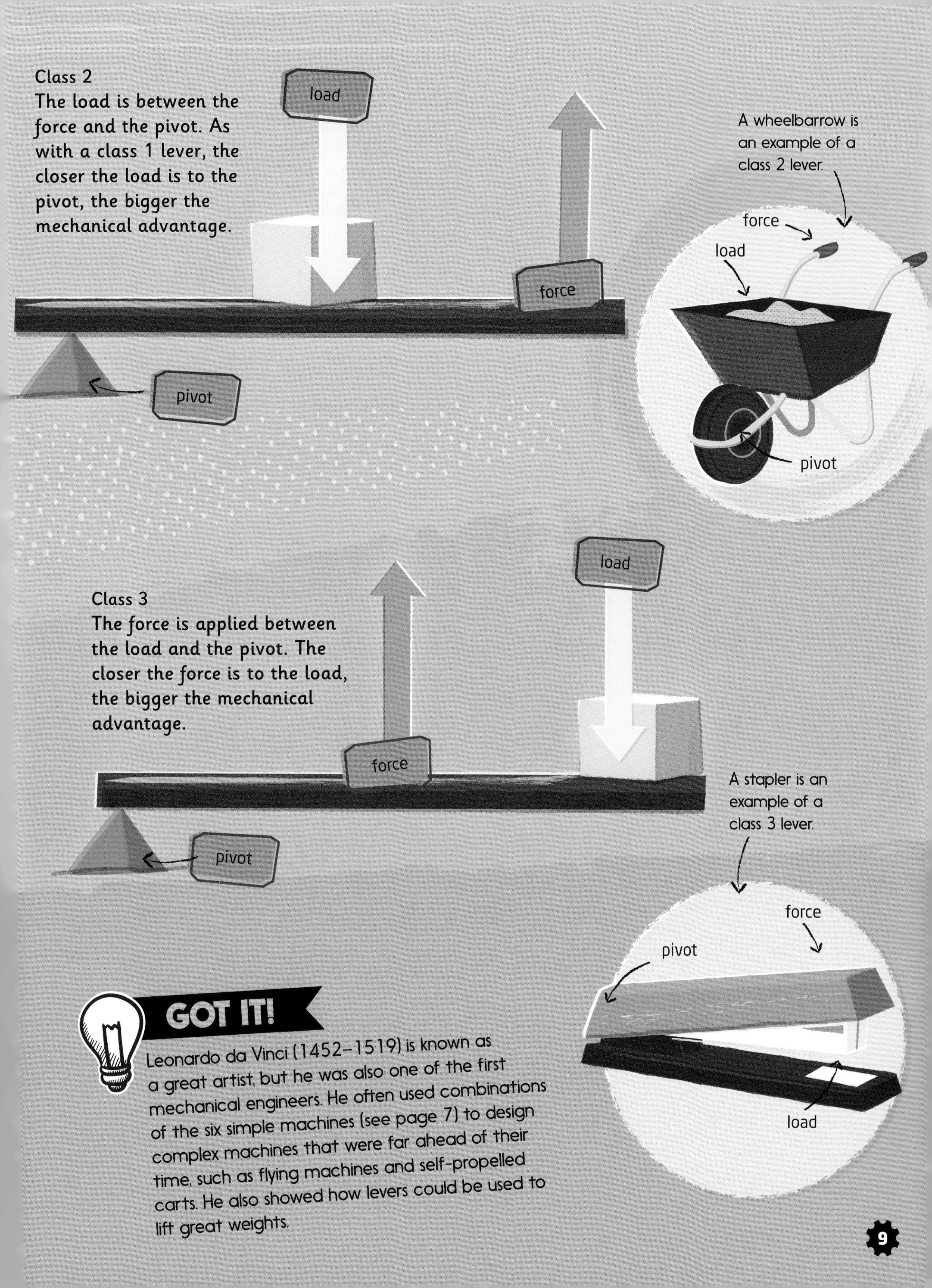
Class 2
The load is between the force and the pivot. As with a class 1 lever, the closer the load is to the pivot, the bigger the mechanical advantage.
load
force
pivot
A wheelbarrow is an example of a class 2 lever.
force
load
pivot
Class 3
The force is applied between the load and the pivot. The closer the force is to the load, the bigger the mechanical advantage.
load
force
pivot
A stapler is an example of a class 3 lever.
force
pivot
load
GOT IT!
Leonardo da Vinci (1452–1519) is known as a great artist, but he was also one of the first mechanical engineers. He often used combinations of the six simple machines (see page 7) to design complex machines that were far ahead of their time, such as flying machines and self-propelled carts. He also showed how levers could be used to lift great weights.

YOU'RE THE ENGINEER: LEVER EXPERIMENTS

Take the first step to becoming a mechanical engineer! Carry out these experiments to better understand levers.

You will need

A wooden ruler
A large binder clip with the metal clips removed
Objects of different weights, such as a pencil sharpener, an eraser, plastic building bricks
A notepad and pencil

1. Balance the ruler on the binder clip (the pivot), so that the clip is exactly in the middle.

2. Place a medium-weight object (the load) at one end. What happens at the other end? Make a note.

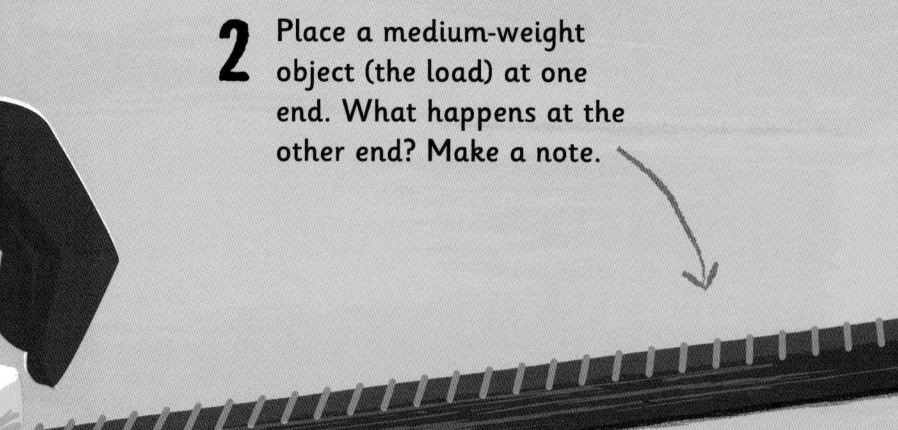

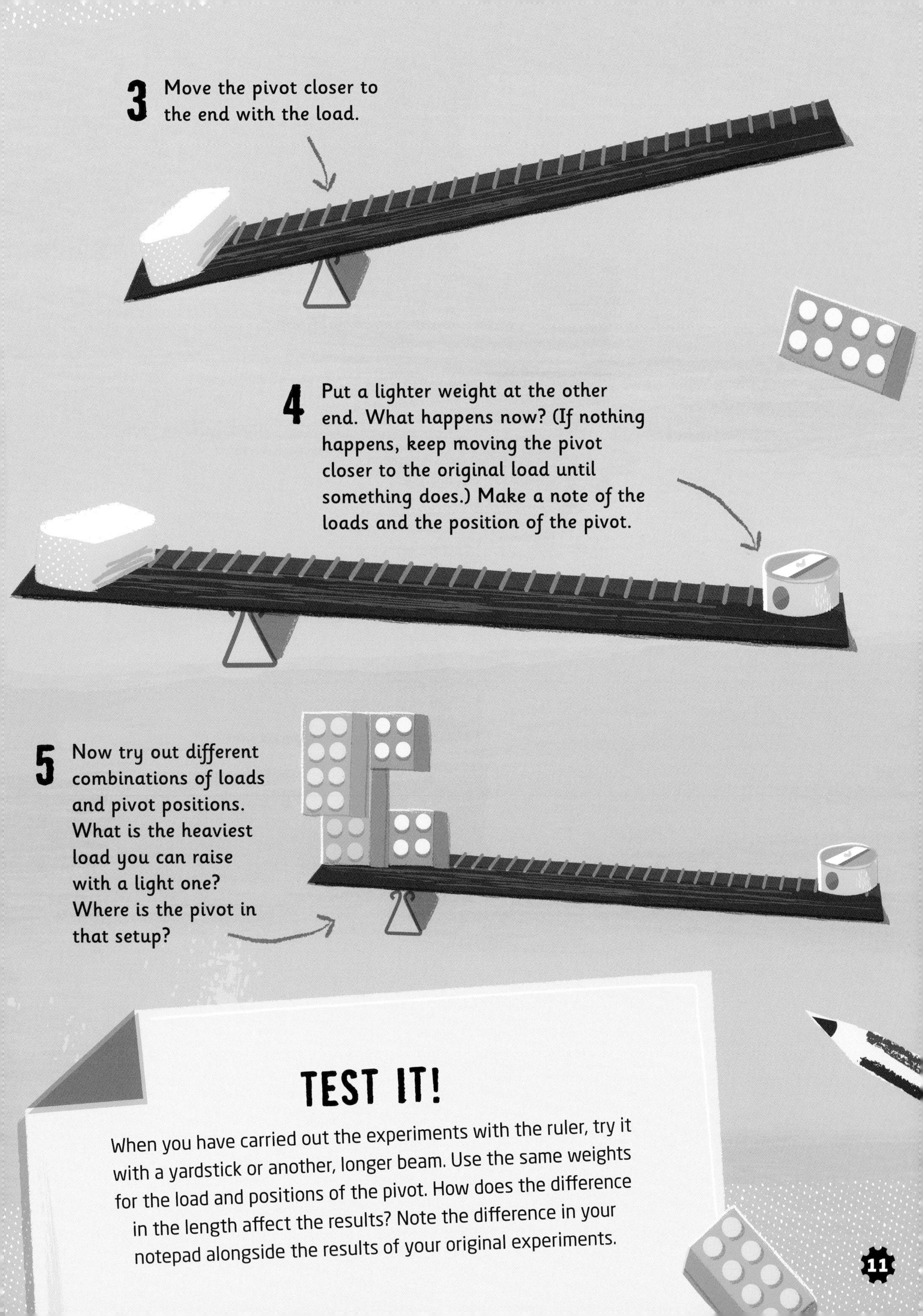

3 Move the pivot closer to the end with the load.

4 Put a lighter weight at the other end. What happens now? (If nothing happens, keep moving the pivot closer to the original load until something does.) Make a note of the loads and the position of the pivot.

5 Now try out different combinations of loads and pivot positions. What is the heaviest load you can raise with a light one? Where is the pivot in that setup?

TEST IT!

When you have carried out the experiments with the ruler, try it with a yardstick or another, longer beam. Use the same weights for the load and positions of the pivot. How does the difference in the length affect the results? Note the difference in your notepad alongside the results of your original experiments.

PULLEYS

Imagine you're a mechanical engineer. You've been asked to design a machine that can lift heavy steel beams ten stories high on a construction site. What simple machine will you start your design with?

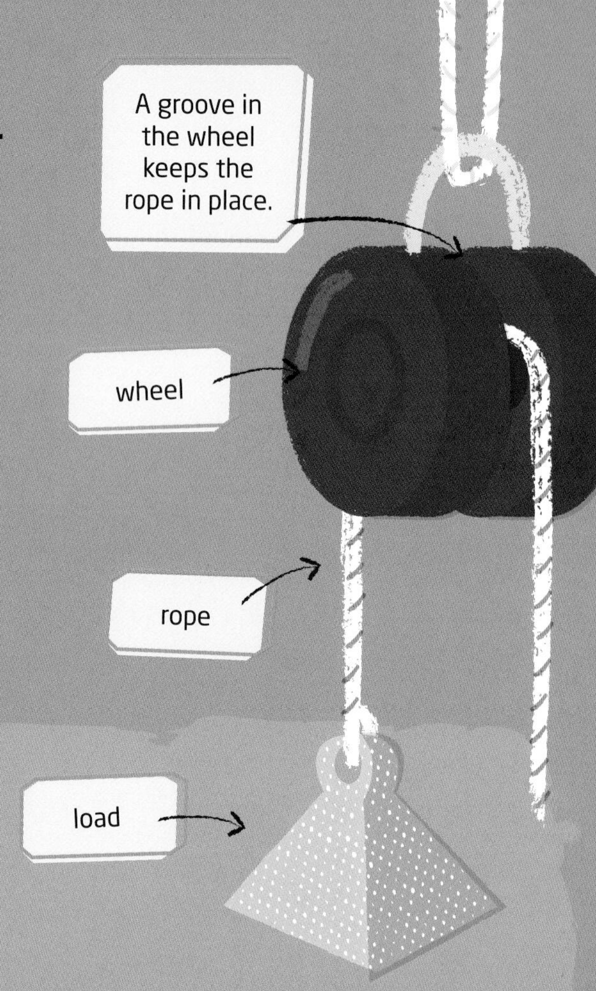

Rope and wheels

A pulley is the perfect solution for lifting and lowering heavy objects. A simple pulley may have just one wheel and rope. More complex pulleys can involve a whole system of ropes and wheels. Pulling down on one end of the rope lifts the load up at the other end.

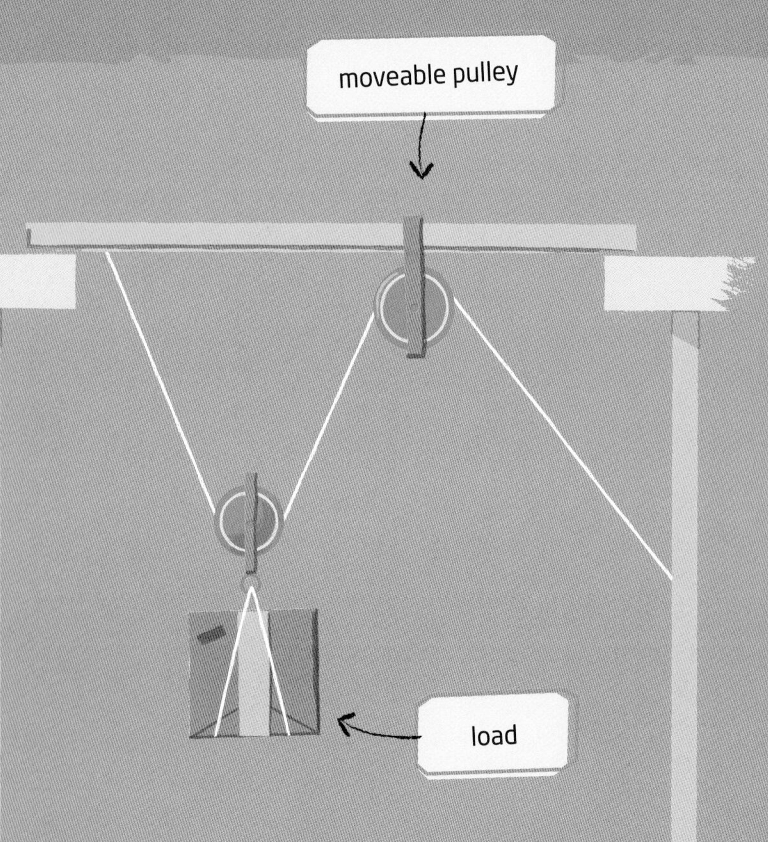

Fixed or moveable?

Some pulleys are fixed. That means the wheel is secured in one place. The flagpole on page 7 is an example of a fixed pulley – the rope raises and lowers the flag to a fixed position. In a moveable pulley, the wheels can move, carrying the load from place to place.

Rising high

One of the most common uses of a pulley system today is in elevators. As buildings get taller and taller, engineers are developing new ways of using pulley systems to lift loads to great heights safely.

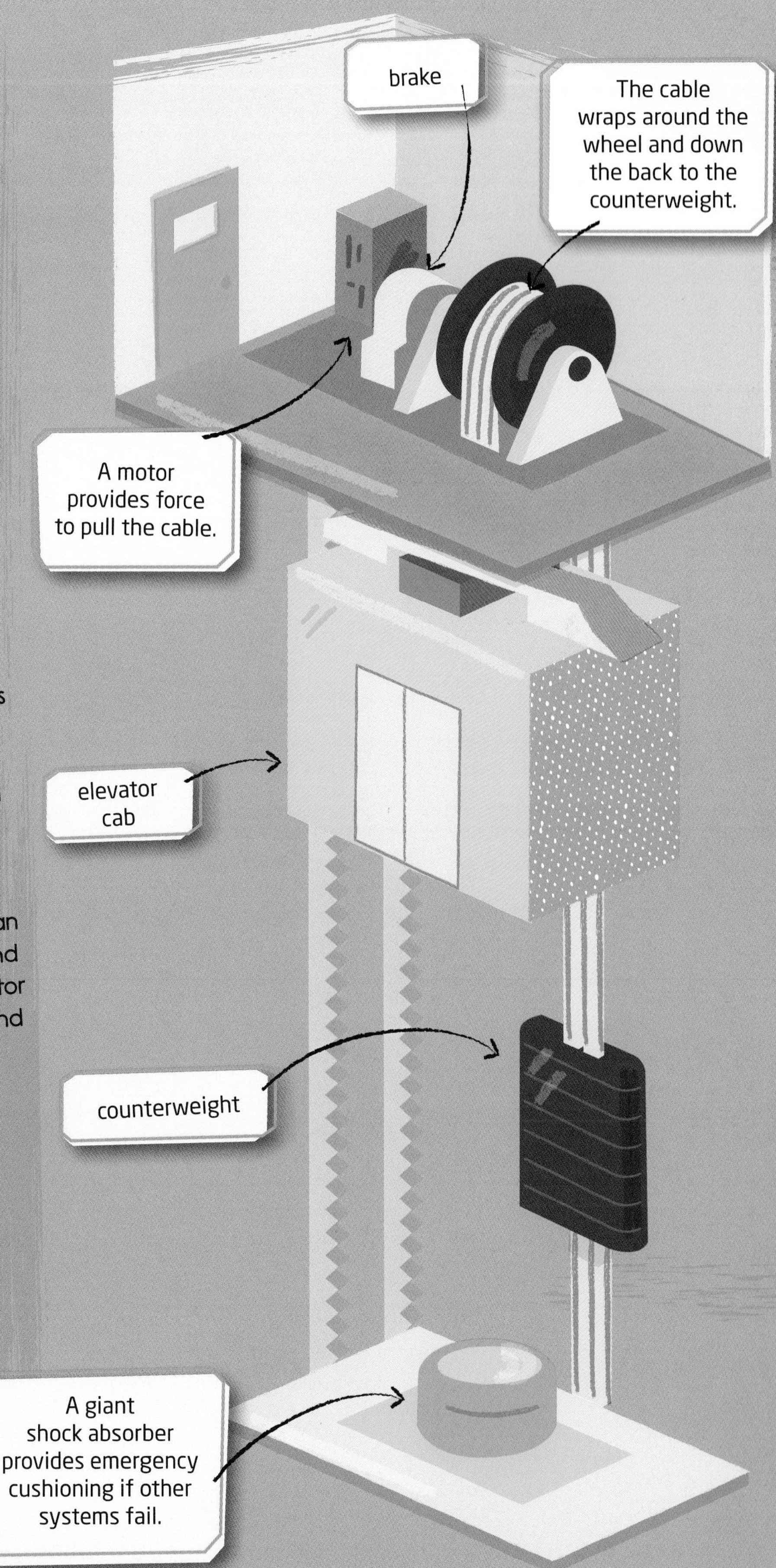

GOT IT!

Before American engineer Elisha Otis (1811–61) came along, people believed it was too dangerous to transport people in elevators using pulleys. What would happen if the elevator cable broke? In 1852, Otis invented a safety device that was attached to the pulley of an elevator. This acted as a kind of brake to stop the elevator from crashing to the ground if the cable snapped.

Using **two wheels** in a **pulley** allows engineers to lift something **twice as heavy** while using the same amount of **input force.**

YOU'RE THE ENGINEER: BUILD A PULLEY SYSTEM

Design and build your own pulley system and see how much weight you can lift up a wall!

You will need

A large sheet of cardboard
Empty tin cans or other light containers (two different sizes)
Strong glue
Masking tape
A length of string
A small bucket
A bottle of water
Pebbles to use as weights
A notepad and pencil

1 Lay the cardboard on the floor. Place the containers in different positions on it. These are going to be the pulley wheels. When you're happy with your arrangement, glue the containers to the cardboard. Attach the cardboard to a wall with masking tape.

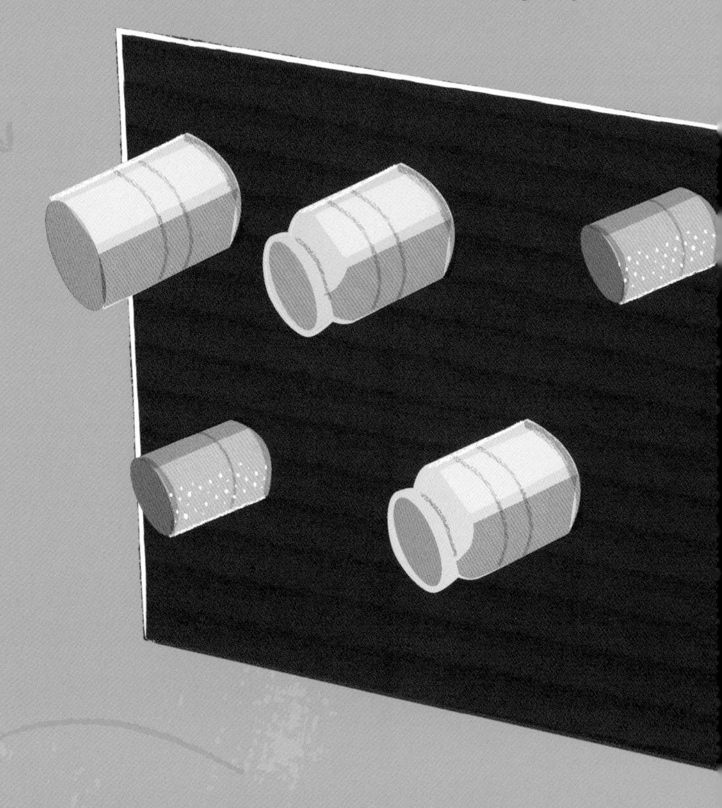

2 Tie the bucket to one end of the string and the bottle of water to the other. The bottle is your load and the bucket is your counterweight.

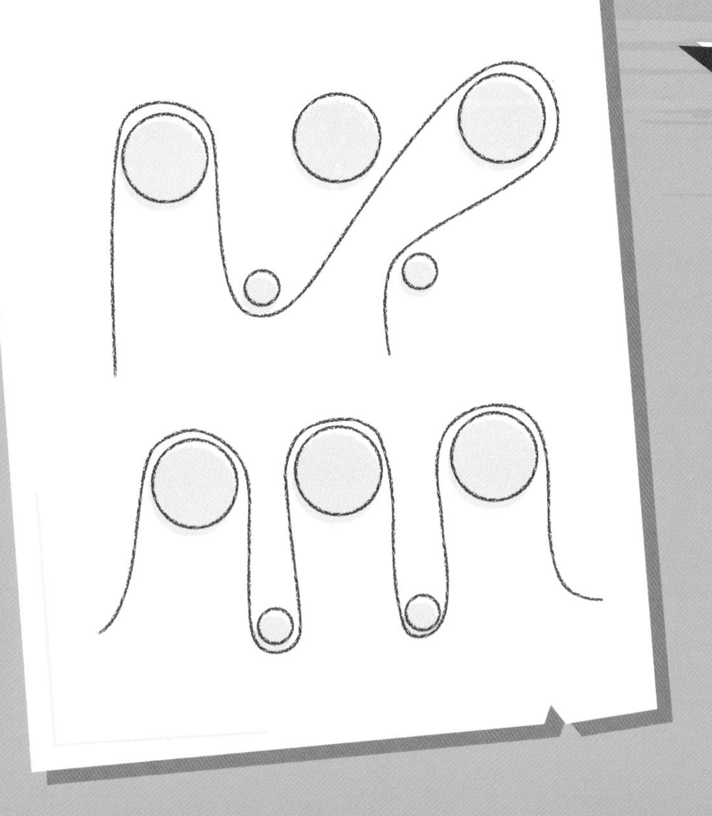

3 On a piece of paper, design two or three different ways that the string could be looped around the pulleys. Decide which one will move the load more easily and why.

4 Next, test your design. Hook the string around the containers according to your first design. Put pebbles in the bucket to lift the bottle. Then try your other designs. Which arrangement allows the bottle to be lifted most easily or with the fewest weights? Is it the one you predicted?

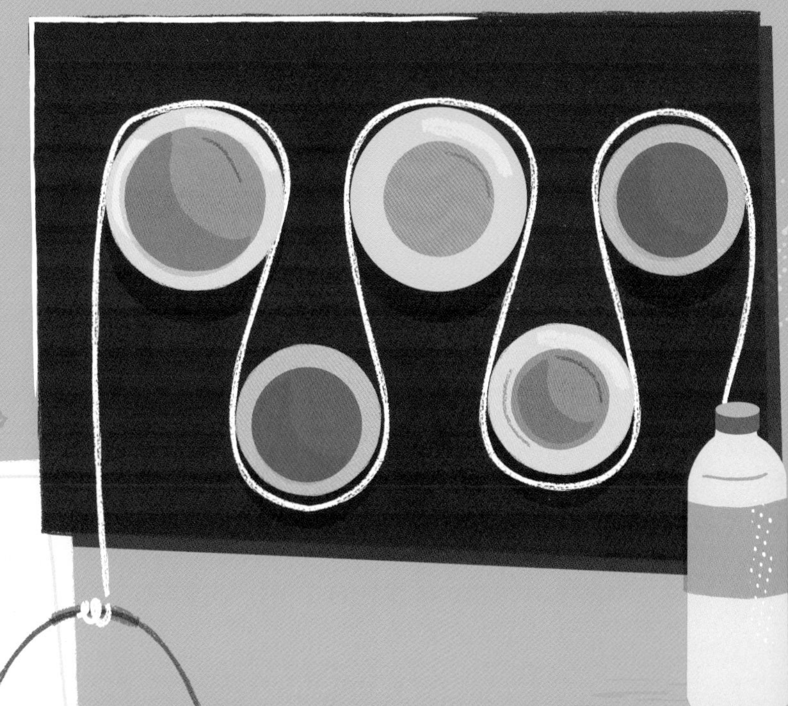

TEST IT!

If you have more containers and cardboard, create a different arrangement of pulley wheels, and try the experiment again. Compare your results with the first system. How does the distance between the wheels affect the force needed to lift the bottle? What effect does changing the amount of water in the bottle have on your experiment?

WHEELS AND AXLES

You probably know that it's easier to move something from one place to another if you put it in something that has wheels. Without wheels, engineers would never have been able to design some of the machines we rely on most.

wheel

axle

How does a wheel work?

A simple wheel is made up of two parts – a large disk (the wheel) with a smaller cylinder (the axle) through the middle of it.

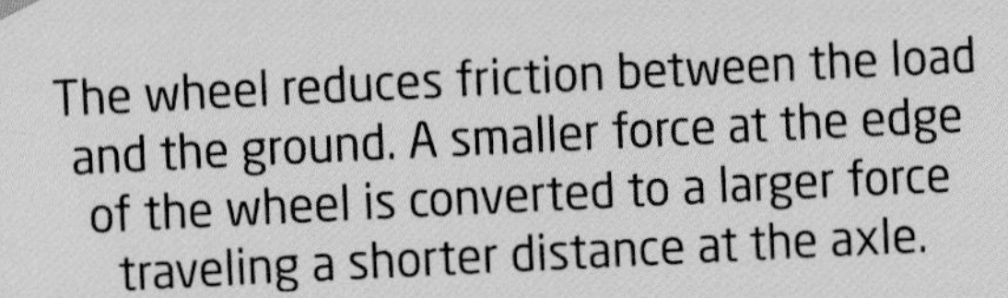

The wheel reduces friction between the load and the ground. A smaller force at the edge of the wheel is converted to a larger force traveling a shorter distance at the axle.

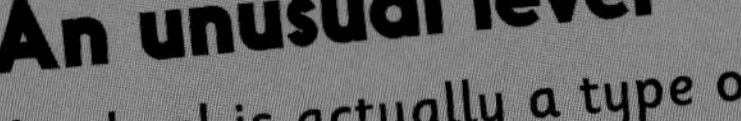

An unusual lever

A wheel is actually a type of lever. It can increase the output force or the distance moved.

A long turn at the edge of the wheel moves the axle with greater force.

A short, strong turn at the axle moves the edge of the wheel a greater distance.

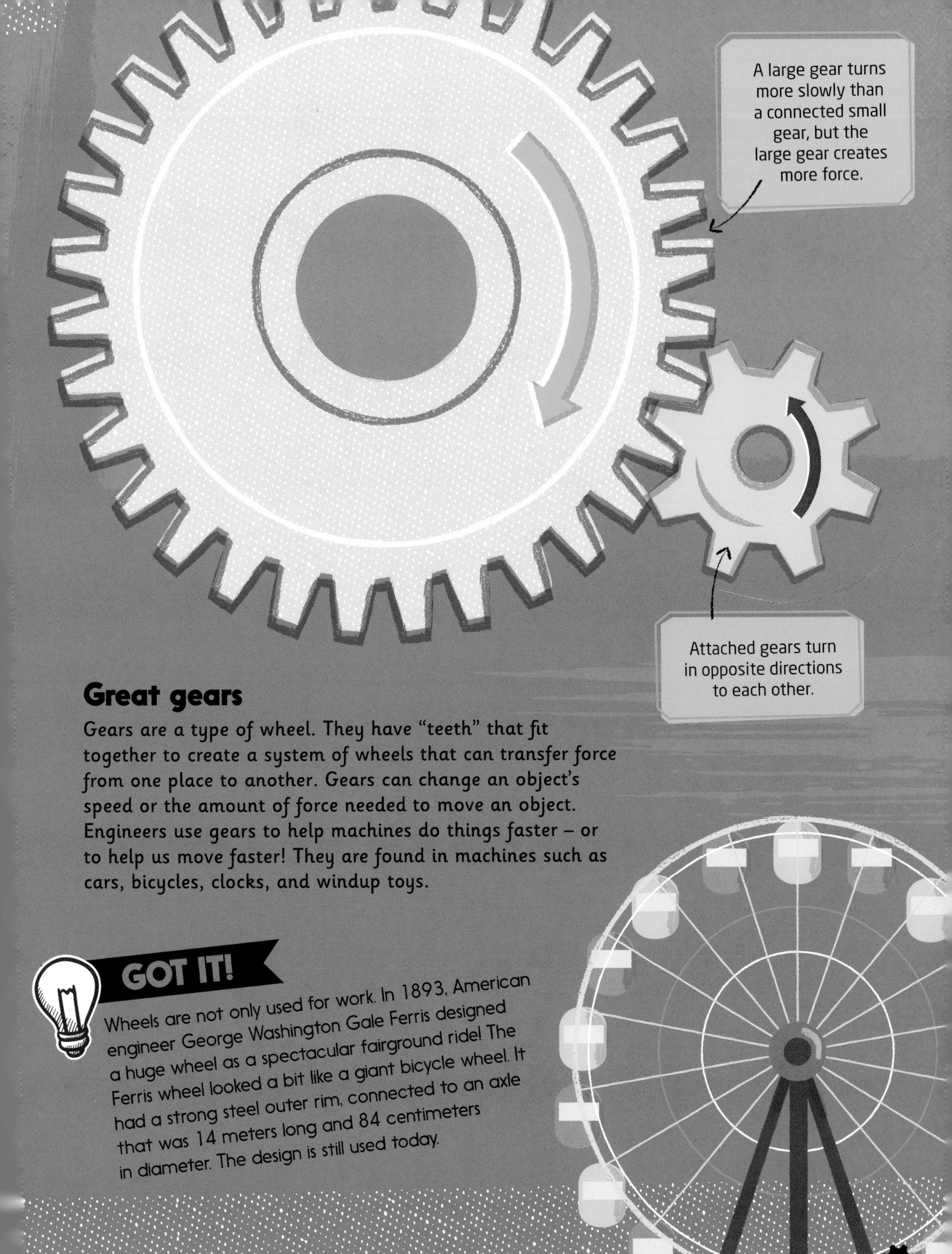

Great gears

Gears are a type of wheel. They have "teeth" that fit together to create a system of wheels that can transfer force from one place to another. Gears can change an object's speed or the amount of force needed to move an object. Engineers use gears to help machines do things faster – or to help us move faster! They are found in machines such as cars, bicycles, clocks, and windup toys.

GOT IT!

Wheels are not only used for work. In 1893, American engineer George Washington Gale Ferris designed a huge wheel as a spectacular fairground ride! The Ferris wheel looked a bit like a giant bicycle wheel. It had a strong steel outer rim, connected to an axle that was 14 meters long and 84 centimeters in diameter. The design is still used today.

YOU'RE THE ENGINEER: WORKING WITH WHEELS

Test how wheels and axles work for yourself by building your own race car out of clothespins and buttons.

You will need

A drinking straw
Scissors
Four wire twist ties
Four buttons of the same size
A clothespin
Colored sticky tape

1 Cut two pieces of the straw, about 2.5 cm long. These are your axles.

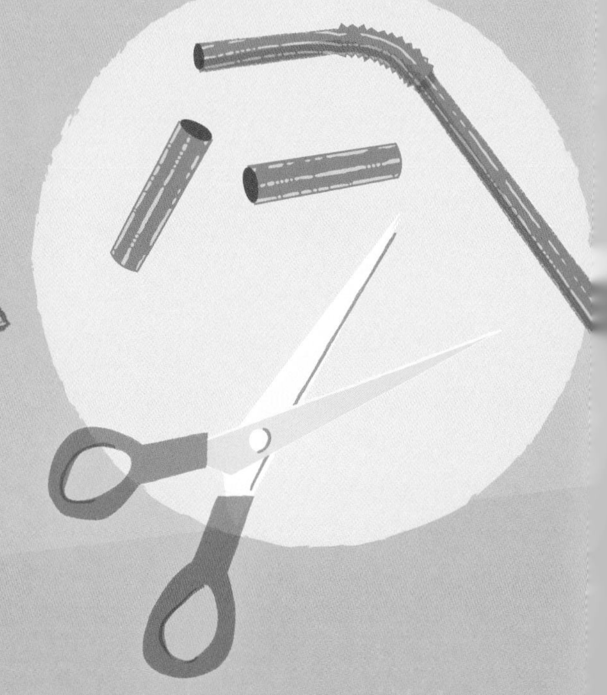

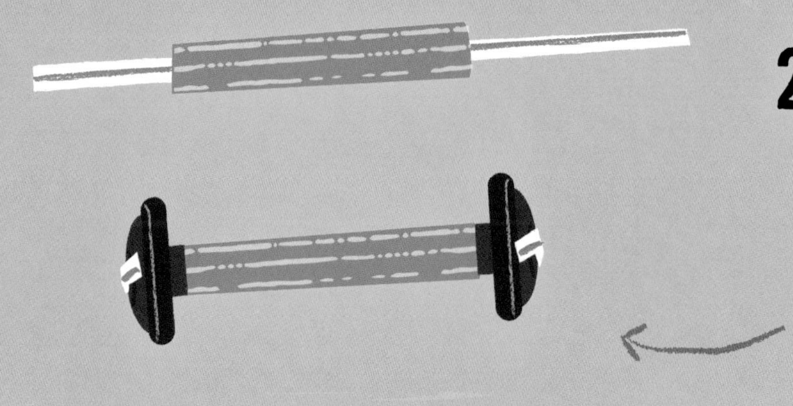

2 Push two twist ties through each of the straws. Attach a button to each end of the twist ties by pushing them through the holes or the loop on the buttons. These are your wheels.

3 Place one axle in the hole at the front of the clothespin. Make sure the wheels are evenly spaced on either side of the clothespin.

4 Slide the other axle in between the clips at the open end. Wind the sticky tape around the clothespin to hold this axle in place.

5 Move your car along the ground and watch how the wheels and axles turn. Try making a second car that has the same design as the first, then race them down a ramp (see page 20).

TEST IT!

Now make one more car, but this time choose two smaller buttons to use for the front wheels. Watch closely as the wheels turn. Compare how many turns the smaller wheels make with how many the larger ones make. What do you notice?

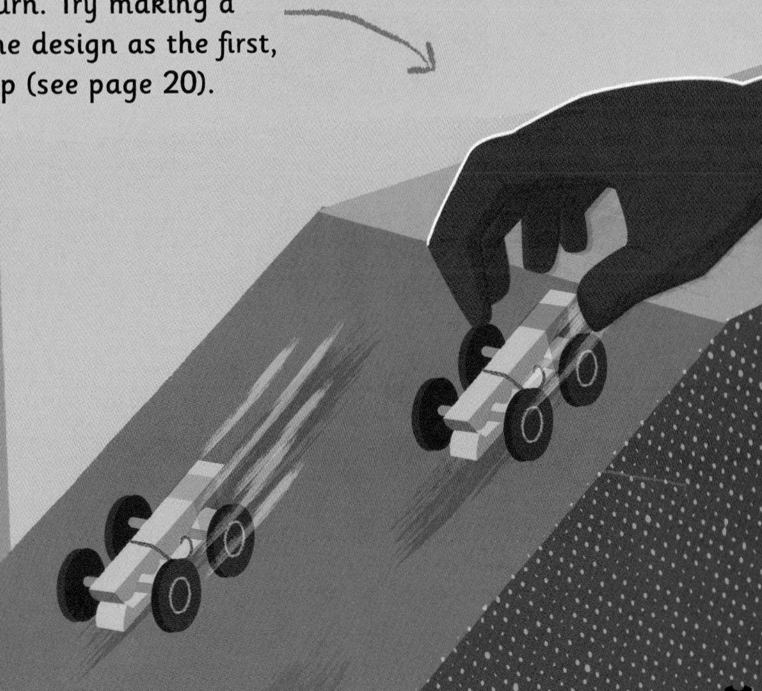

INCLINED PLANES AND WEDGES

An inclined plane (usually known as a ramp) is a flat surface with one end higher than the other. That doesn't seem like much of a machine, does it? But for mechanical engineers, ramps are key to many useful machines.

High and low

A low ramp requires less input force, but the object travels a greater distance to reach the same height. A steep ramp requires more input force, but the distance traveled to reach the same height is less.

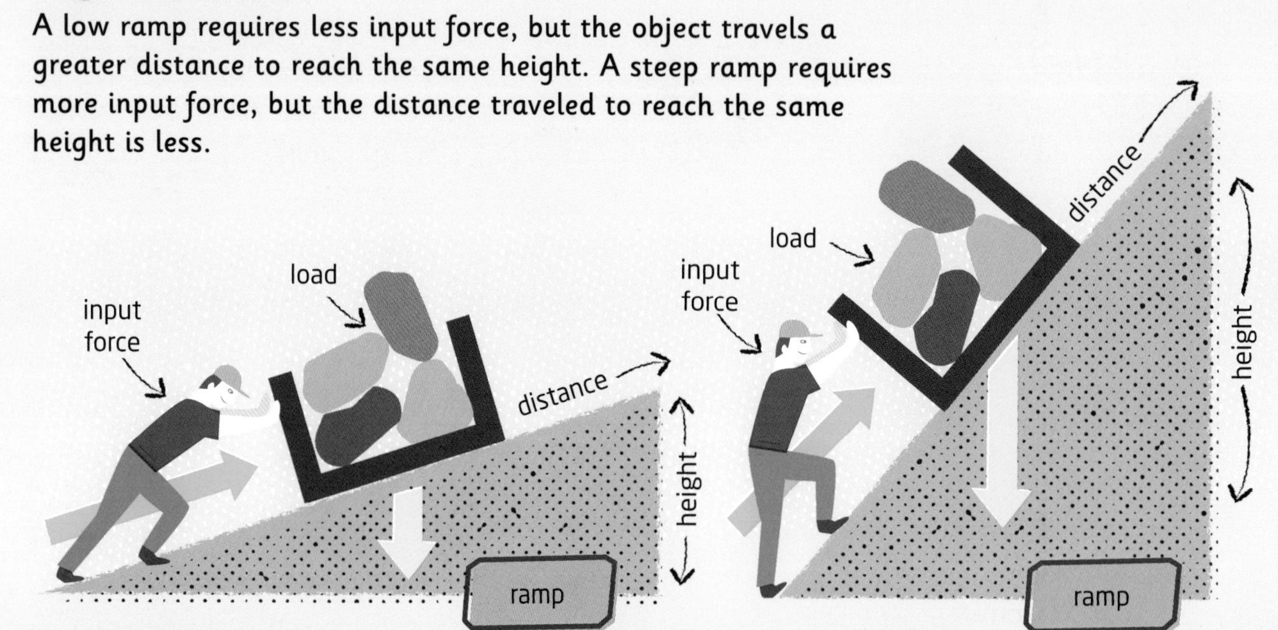

Ramp it up!

Ramps are used in many different complex machines. How could cars get onto a car carrier without a specially engineered ramp? These are often raised and lowered by an engineering system called hydraulics.

Slide to safety

Have you ever seen an airplane evacuation slide? These ramps are engineered using materials such as neoprene, which is flame-retardant. This helps people to escape safely in an emergency.

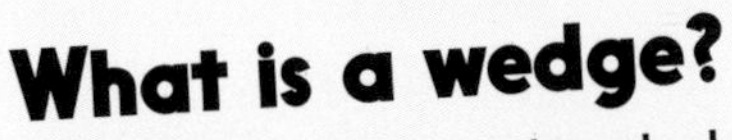

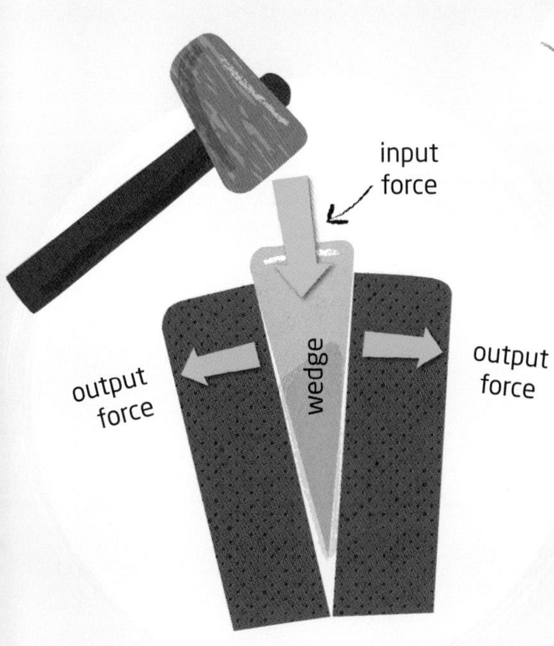

What is a wedge?

A wedge is two inclined planes that have been put together to make a tool that is thin at one end and thick at the other. A wedge turns a small input force at the thick end into a large output force at the thin end.

Wedges are often used for splitting things apart. For example, axes, knives, and nails are all types of wedges.

GOT IT!

The great pyramids of Egypt are an amazing feat of engineering, especially when you think that they were built around 4,500 years ago. But without cranes or any other mechanical machinery, how did the ancient Egyptians build the pyramids so high? Experts think they might have used ramps to lift the stones. These may have been built in a spiral around the pyramid as it got higher.

Supersonic **jets,** race **cars,** and **speedboats** all go **fast** because of their **wedge-shaped fronts.** These "split" the air or water and **carve** a fast path.

SCREWS

A screw is an inclined plane wrapped around a cylinder, rod, or cone in a spiral pattern. In a screw, the path of the plane is the grooves or threads. Engineers often use screws for lifting things – even water!

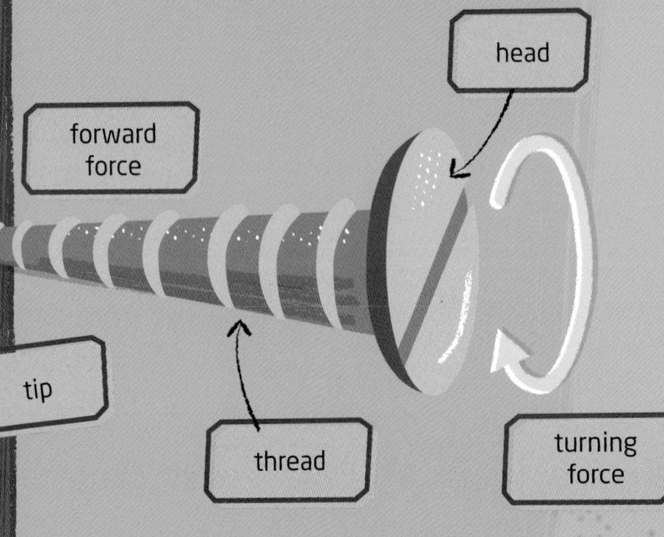

How do screws help?

Think about an actual screw. As you turn it, the thread forces the screw forward into the wood or other material. The force is spread out along the whole thread.

Everyday screws

Screws don't just hold doors on their hinges and keep pieces of furniture together. Every time you turn on a tap, unscrew the lid of a water bottle, or screw in a light bulb, you're using a screw!

An engineering essential

Screws hold machines together, but they can also be the main part of the machine itself. For example, drills are huge screws that are used for digging up roads or drilling for oil.

GOT IT!

Archimedes (c. 287–212 BCE) was an ancient Greek engineer and mathematician. He designed a machine that could raise water. It was made up of a screw tightly fitted inside a cylinder, with a handle at one end. The other end was put in the water. Turning the handle twisted the screw, which raised the water up the threads. Archimedes' screw was soon used to raise water from rivers to irrigate crop fields. It is still used in parts of the world today.

YOU'RE THE ENGINEER: ARCHIMEDES' SCREW

See for yourself how to raise water using a simple screw by building your own version of Archimedes' amazing machine.

You will need

A plastic tube about 3 cm in diameter
Thin, flexible tubing
Clear tape
A bowl
Food coloring
A glass

1 Take the plastic tube and wrap the thin tubing around it in a spiral from top to bottom. Use the clear tape to hold it in place.

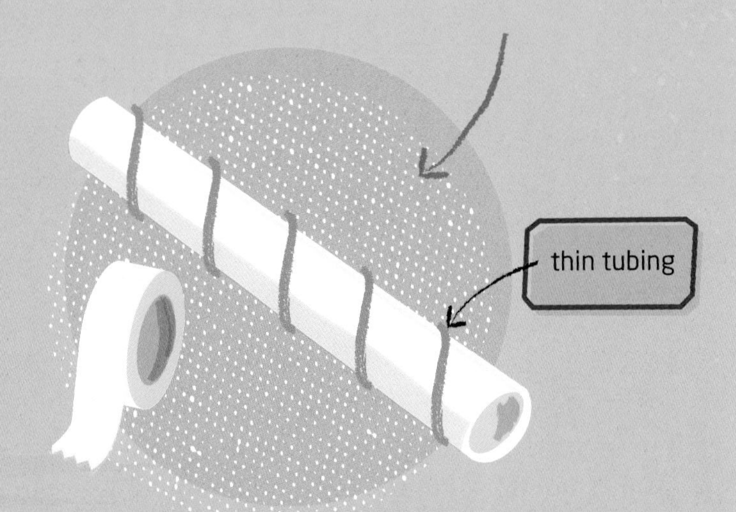

2 Pour water into the bowl until it's about half full, then add a few drops of food coloring.

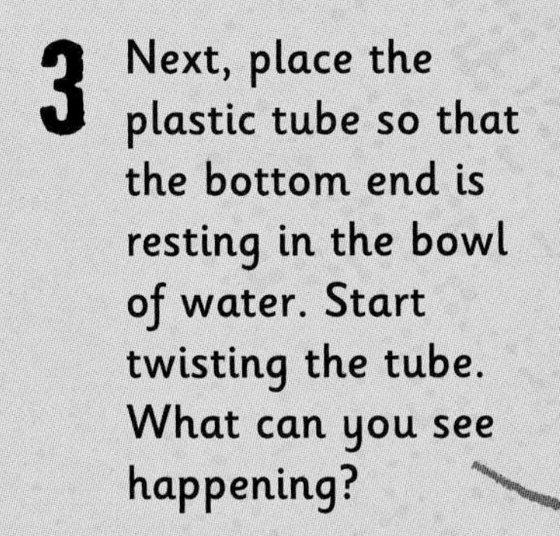

3 Next, place the plastic tube so that the bottom end is resting in the bowl of water. Start twisting the tube. What can you see happening?

4 As you turn the tube, the water will rise through the thin tubing. Make sure you catch it in the glass as it comes out of the top!

TEST IT!

Try raising and lowering the screw. How do different gradients affect the raising of the water?

Try turning the screw faster and slower. What happens to the water in the screw for each difference in speed?

What do you notice about where the water is in the spiral tube?

COMPLEX MACHINES

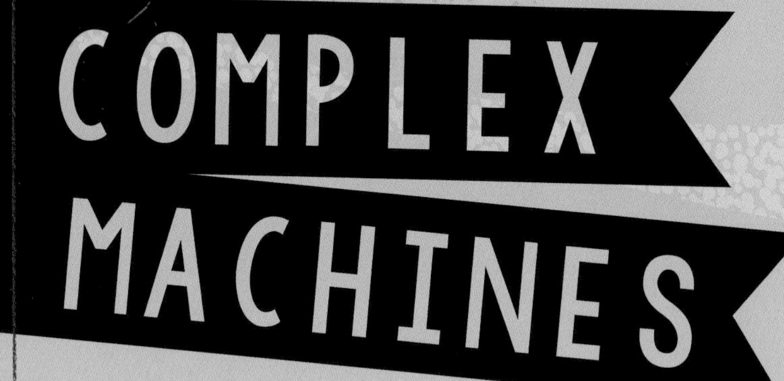

On their own, simple machines can help us in many ways. But when engineers start combining simple machines to make more complex machines, the possibilities are endless.

Bicycle engineering

A bicycle is a good example of how several simple machines can be combined to make up a complex machine.

Screws are used to hold different parts of the bicycle together.

The pedals are a type of lever – when they move up and down, they work the pulley.

Chain and gears form a pulley system.

wheel and axle

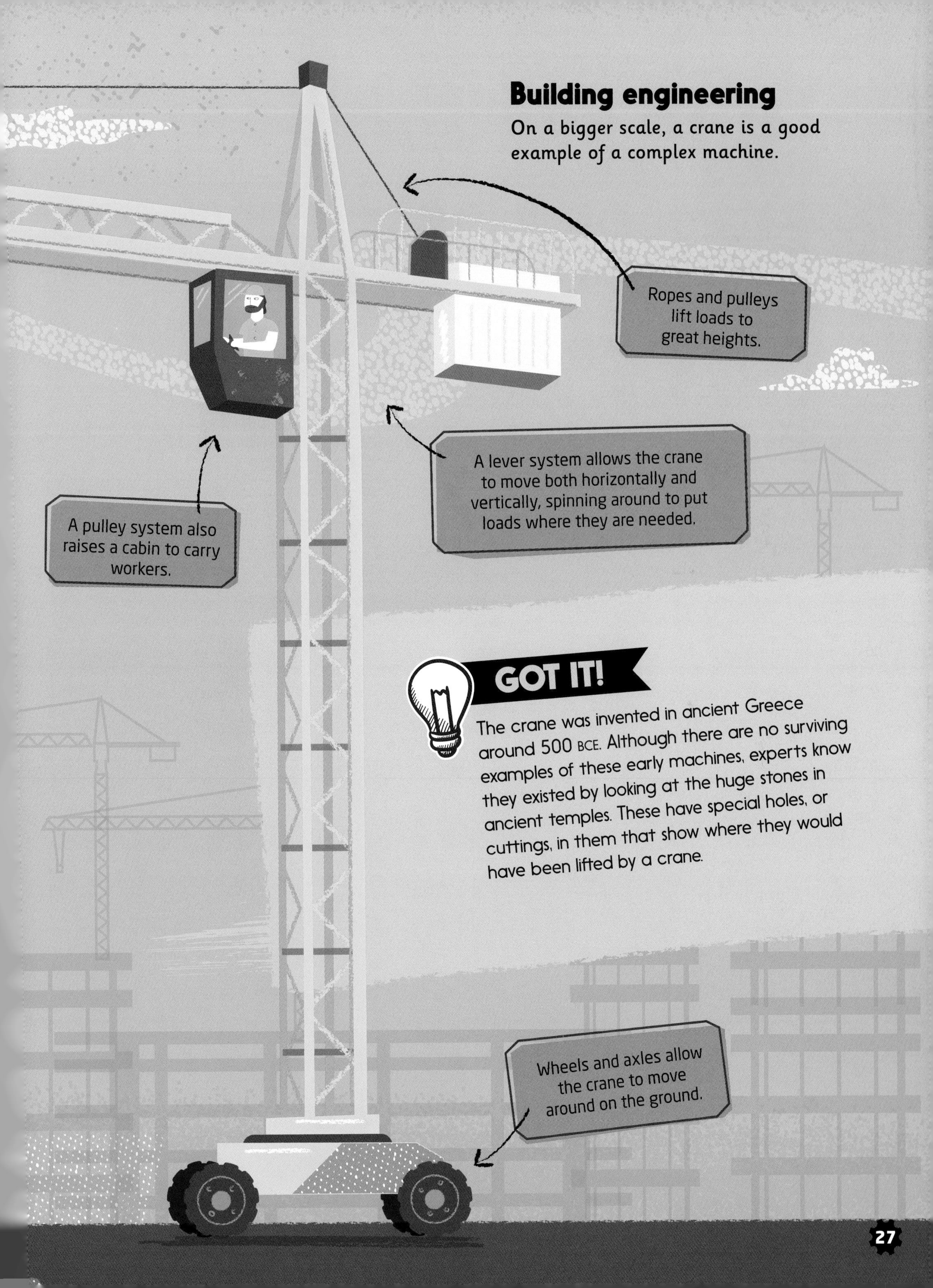

Building engineering

On a bigger scale, a crane is a good example of a complex machine.

GOT IT!

The crane was invented in ancient Greece around 500 BCE. Although there are no surviving examples of these early machines, experts know they existed by looking at the huge stones in ancient temples. These have special holes, or cuttings, in them that show where they would have been lifted by a crane.

POWERFUL MACHINES

One of the most useful things mechanical engineers work on is designing machines to generate power.

Turbine talk

A turbine is any device that converts energy from moving fluid or air into mechanical energy. A turbine could be...

...a waterwheel, where water turns a wheel that powers machines often used in mills...

...or a jet engine, which sucks in air that is compressed by the blades (wheel) and then set alight to power the engine.

In this wind turbine, the force is applied in the middle (axle) and is increased in the blades (wheel) over a greater distance.

The power of combustion

Combustion engines are power-producing machines used in cars and planes. In the engine, pistons mix fuel and air, which is then burned to create energy. The engine spins the axle connected to the wheels, making them turn and the vehicle move.

A combustion engine combines machines such as a lever and a wheel and axle.

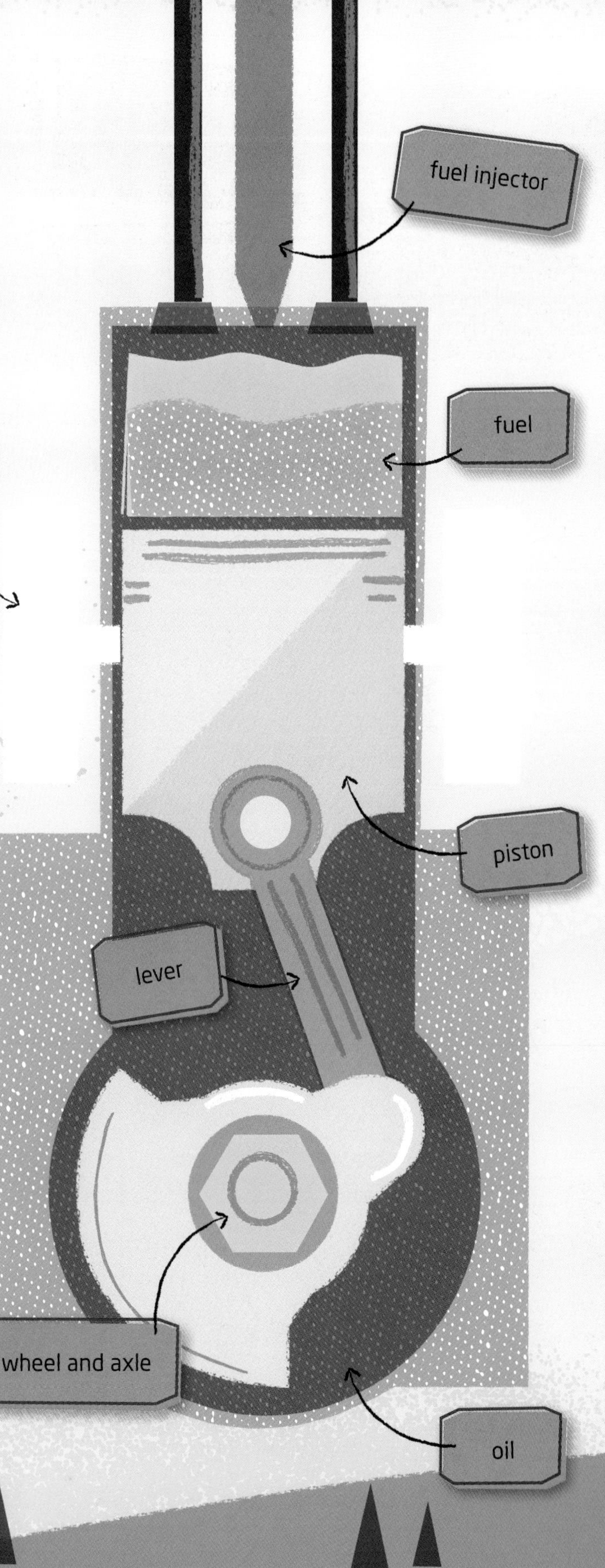

Be a mechanical engineer!

Choose two or three of the simple machines you have read about in this book. Think of a way you could combine them to invent a new machine.

1 First identify a need. What are you trying to do? For example, are you trying to move something heavy across a flat surface, or lift it up?

2 Draw some sketches to show how your idea might work. Think about what shapes and materials the different parts should be.

3 If you can, build a model of your idea and try it out. What works and what doesn't? Think about how to improve your idea.

GLOSSARY

axle the cylinder that goes through the middle of a wheel

complex describing something that has several different, connected parts

counterweight a weight used in a pulley to balance the weight of the load

diameter the width of a circle from one side to the other

equation a mathematical sentence containing two sides separated by an equal sign (=)

force the amount of energy applied to an object

friction the force that acts between two surfaces when they move past each other

gear a type of wheel which has "teeth" that interlock with other gears

gradient how steep or shallow an incline (slope) is

hydraulics an engineering system that uses water to perform mechanical tasks

input force the force a person applies to a machine

magnitude a measure of the force applied to an object

mechanical advantage the amount of work that a machine saves

output force the force a machine applies to an object

pivot the part of a lever that the beam balances on and moves around

renewable resources sources of energy that will never run out, such as wind, water, and the sun

work the amount of energy it takes to move an object over a distance

INDEX

COLLECT THEM ALL!

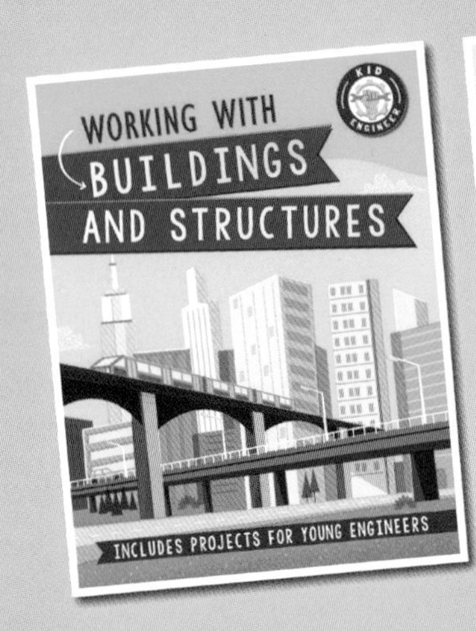